环保卫士
小猪爱可

什么是"绿色生活"

小猪爱可教你"绿色生活"

[美] 丽莎·弗兰奇 著　　[美] 巴瑞·戈特 绘　　张玉亮 译

江西科学技术出版社

Original title: ECO-Pig Explains Living Green: What Does It Mean to Be Green

Copyright © 2010 by Abdo Consulting Group, Inc. International copyrights reserved in all countries.

All rights reserved. No part of this book may be reproduced in any form without written permission from the publisher.

The simplified Chinese translation rights arranged through Rightol Media

（本书中文简体版权经由锐拓传媒取得Email:copyright@rightol.com）

版权合同登记号 / 14-2016-0009

图书在版编目（ＣＩＰ）数据

什么是"绿色生活"：小猪爱可教你"绿色生活"：英汉对照 / (美) 弗兰奇著；(美) 戈特绘；张玉亮译.
-- 南昌：江西科学技术出版社, 2016.8
（小猪爱可讲环保）
ISBN 978-7-5390-5488-9

Ⅰ.①什⋯ Ⅱ.①弗⋯ ②戈⋯ ③张⋯ Ⅲ.①环境保护－少儿读物－英、汉 Ⅳ.①X-49

中国版本图书馆CIP数据核字(2016)第026836号

国际互联网（Internet）地址：http://www.jxkjcbs.com
选题序号：KX2016083　　图书代码：D16001-101

小猪爱可讲环保
什么是"绿色生活"：小猪爱可教你"绿色生活"

文 /(美)丽莎·弗兰奇　　图 /(美)巴瑞·戈特　　译 / 张玉亮
责任编辑 / 刘丽婷　　美术编辑 / 刘小萍　曹弟姐
出版发行 / 江西科学技术出版社
社址 / 南昌市蓼洲街2号附1号　　邮编 / 330009
电话 / (0791)86623491　　86639342(传真)
印刷 / 江西华奥印务有限责任公司
经销 / 各地新华书店
成品尺寸 / 235mm×205mm　　1/16
字数 / 200千　　　　印张 / 8
版次 / 2016年8月第1版　2016年8月第1次印刷
书号 / ISBN 978-7-5390-5488-9
定价 / 50.00元（全4册）

赣版权登字-03-2016-7　　版权所有，侵权必究
（赣科版图书凡属印装错误，可向承印厂调换）

Dear friends of the earth,
I just have to insist
that you pay close attention
to this tale with a twist.

地球上的各位朋友们，我希望大家带着思考仔细读一读下面的故事。

It belongs to a pig
who hangs out in a tree.
His name is Bernard,
but we call him E.P.

　　有一只猪喜欢在树上玩
耍，他叫佰纳德。但是大家都
叫他"E.P."。

The *E* stands for eco,
short for ecology.
That means he cares for our planet.
Listen up, and you'll see.

"E"代表生态、环保。因为他时
刻关心着我们所在的地球。好好听下面
的故事，你就会明白了。

One moonlit fall night
in a town called To-Be,
strange noises were heard
in a green apple tree.

　一个秋天的夜晚，月光洒满未来
小镇，绿色的苹果树上传来阵阵奇怪
的声音。

6

What was making that noise?
Now, what could it be,
that snorting and snuffling
in the green apple tree?

是谁发出的这些声音？到底
会是谁在绿色的苹果树上发出这
种"呼哧呼哧"的喘气声？

By the light of the moon
the To-Bes gathered about,
when through the shivering leaves
out poked a pig's snout.

　　未来小镇的居民们借着月色都聚集到苹果树下来看个究竟。只见一只猪鼻子从沙沙作响的树叶中悄悄伸了出来。

"Good golly, gracious,
there's a pig in our tree!
Why has he come?"
asked Grandpa To-Be.

"哎呀，天呐，我们的树上
有一头猪！他怎么跑到这儿来的
呢？"一位老爷爷问。

"I have come," said E.P.,
"because my snout led me here!
Once I sniffed your clean air,
there was nothing to fear.

"是我的鼻子指引着我来到这里的。"小猪爱可说，"当我呼吸到这里的清新空气，我就变得无所畏惧啦。

"Then I sniffed your fresh flowers,
your mountains, your streams,
your forests, and your pastures.
It's the town of my dreams!

"然后我又闻到了你们这里的鲜
花、青山、溪流、森林，还有草地的
芳香。这里就是我的梦想家园啊！

11

"But something is wrong here,
in this grand eco scene.
When I start to sneeze,
that means someone's not Green!"

　"但在这一派欣欣向荣的生态景象中，好像有什么地方不大对劲。如果我开始打喷嚏的话，那就意味着有人还不够'绿色环保'！"

Tiny To-Be looked at her hands,
her feet, and her knees.
"We do come in all colors,
more than just green, if you please."

一位小朋友满脸疑惑地看完自己的
手、脚和膝盖后，对小猪爱可说："拜托
你仔细看看，我们有各种颜色，不仅仅是
绿色呀。"

"When you are Green," said E.P.,
"that just means that you care,
for Earth, this great planet,
our home that we share.

"当你'绿色环保'时，"小猪爱可说，"这就意味着你关心地球——这颗伟大的星球，关心我们共同生活的家园。

14

"If we live a little less large
and a little more lean,
we'll help keep our home cool
and we'll help keep it clean."

"如果我们住得小一点儿，
生活得简朴一点儿，就可以让我
们的家更凉爽，更干净。"

15

"Being Green sounds too hard!"
sighed Teen To-Be.
"It's quite easy," said Lou,
"and you can learn from E.P."

　　"保持'绿色环保'听起来好难啊！"一位小哥哥叹息道。"其实很简单。"山羊洛尔说，"你们可以向小猪爱可学习。"

"First, please get me a tissue
and a nice cup of tea.
My snout has an issue
with Mr. Watts of To-Be.

"首先，请给我一张纸巾和一杯好
茶。我的鼻子有点儿不习惯未来小镇的
瓦特先生。

"Watts glows in the dark
and he glows in the day.
His whole house is turned on,
so I have to say 'Hey!'

"瓦特家日夜通明。他家里的所有电器
都一直开着，我必须得跟他说'停'了！

"When we plug in too much
we burn too much fuel,
which pollutes planet Earth.
Now we know that's not cool!

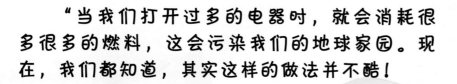

"当我们打开过多的电器时，就会消耗很多很多的燃料，这会污染我们的地球家园。现在，我们都知道，其实这样的做法并不酷！

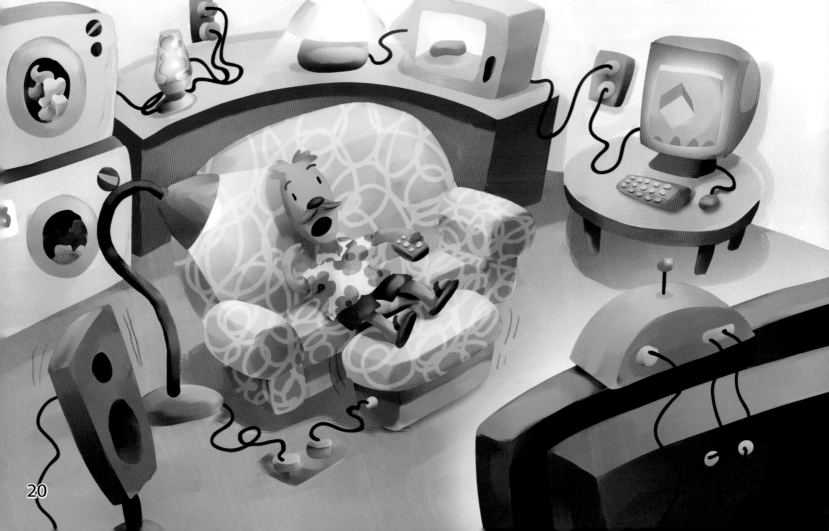

20

"So turn it off and unplug it.
Don't use more than you need.
Just use your fair share.
That's the way to succeed!"

"现在，让我们把多余电器的电源关闭并且拔下插头吧。不要贪多，足够用就好。这才是正确的做法！"

21

But just at that moment
the Guzzlers drove by,
trailing exhaust fumes.
Oh my goodness, oh my!

就在此时，耗油大户——古茨勒
一家开着车经过，排出一大串尾气。
天啊，我的天啊！

They drove round and round,
and E.P. started to wheeze.
Whoa! Let's step on the brakes
and get them bicycles please.

他们驾驶着车四处溜达。小猪爱可开始
气喘吁吁。咳咳咳！小猪爱可好想说：你们
赶紧踩下刹车，换骑自行车兜风吧。

23

You can ride bikes to the store,
to the park, and to the school.
So let's ride and breathe easy.
It's an Eco-Pig rule!

你们可以骑自行车去商场、去公园、去学校。让我们一起骑车吧，这样可以让呼吸更舒畅呢。这就是小猪爱可的环保规则哦！

They all grinned as they rode
to see the Heaps of To-Be.
There, they learned that recycling
is fun, easy, and free!

当他们骑车看到未来回收站时，他们都开怀大笑着。在那里，他们了解到废物回收非常有趣，也很简单，而且还免费呢！

The Heaps recycle it all,
old clothes and old shoes,
cans, plastic, and glass,
and yesterday's news.

回收站对旧衣物和旧鞋、易
拉罐、塑料、玻璃和过期的报纸
进行充分的回收利用。

26

Recycle the rainwater.
Don't let it go down the drain.
Water your grass and your garden.
Wash your sheepdog Duane.

回收利用雨水不要让它浪费掉。你可以用它浇灌你的草坪和花园，用它给你的牧羊犬杜安洗澡。

"If you live Green," said E.P.,
"you can be sure of one thing:
our planet will thank you.
You will hear her heart sing,

"如果你按照'绿色'环保的方式生活，"小猪爱可说，"你至少可以肯定一件事：我们所居住的星球一定会感谢你的。你都能听到来自她内心最欢快的歌声。

in the wind and the stars,
the bluebirds and the bees,
in you and in me,
and the green apple trees!"

"在这青翠的苹果树上，在清风和
闪烁的星星中，在蓝色知更鸟和蜜蜂
的甜美歌声中，在你和我身上，你都
能感受到她的开心和喜悦。"

29

必学词汇

生态学——研究动植物与它们所处环境间关系的一门科学。

废气——发动机排出的气和烟。

绿色环保——与环境或保护环境相关的（事物）。

污染——用人造废弃物破坏环境的行为。

回收利用——将废物、玻璃或易拉罐分类回收，以便再次利用。

你知道吗？

- 回收利用1吨纸，可节约：17棵树、26.3立方米水、1.8立方米石油以及4077千瓦时的电力；或者可以减少：266.3千克空气污染、2.3立方米垃圾填埋空间。

- 美国人平均每年使用295千克纸张。如果我们回收利用所有纸张，就可节约1亿吨的木材。

- 美国人每小时消耗250万个塑料瓶。

- 回收利用的塑料可重新制成很多东西，包括木塑复合地板、衣物、花盆、睡袋和滑雪服的隔热材料，以及车挡。

- 回收利用一个铝制易拉罐所节约的电力足够供一台电视运转3小时。

- 回收3900万台电器所获得的钢铁足够建造约160个足球场。

实现绿色环保的更多方法

与你的父母聊聊
你们在家中可以做些什么

1. 将供热装置调低5℃。

2. 关掉不需要的灯。

3. 用风扇代替空调来降低室温。

4. 最好用淋浴洗澡而不是用浴缸泡澡。

5. 使用节能灯泡。

6. 手工洗碗，不要使用洗碗机。

7. 避免使用一次性用品，如餐巾纸。

8. 刷牙和洗手时不要让水哗哗白流。

9. 检查水龙头有没有渗漏。

10. 开辟一片菜园。